Henry Macnaughton-Jones

Medical Responsibility in the Choice of Anaesthetics

Henry Macnaughton-Jones

Medical Responsibility in the Choice of Anaesthetics

ISBN/EAN: 9783337170868

Printed in Europe, USA, Canada, Australia, Japan

Cover: Foto ©berggeist007 / pixelio.de

More available books at **www.hansebooks.com**

MEDICAL RESPONSIBILITY

IN THE

CHOICE OF ANÆSTHETICS

WITH A

TABLE OF THE ANÆSTHETIC EMPLOYED, ITS MODE OF ADMINISTRATION
AND RESULTS, IN NEARLY FIFTY LARGE HOSPITALS IN THE
UNITED KINGDOM.

BY

H. MACNAUGHTON JONES, M.D., M.CH.

Fellow of the Royal Colleges of Surgeons Ireland and Edinburgh;
Surgeon, Cork Ophthalmic and Aural Hospital;
Physician, Cork Fever Hospital, Cork Maternity, and Children's Hospital;
Late Senior Demonstrator (Lecturer on Surgical and Descriptive Anatomy) Queen's College, Cork;
President Elect, South of Ireland Branch of the British Medical Association.

LONDON:
H. K. LEWIS, 136 GOWER STREET.
DUBLIN; FANNIN & CO., GRAFTON STREET.
1876.

MEDICAL RESPONSIBILITY

IN THE

CHOICE OF ANÆSTHETICS

WITH A

TABLE OF THE ANÆSTHETIC EMPLOYED, ITS MODE OF ADMINISTRATION
AND RESULTS, IN NEARLY FIFTY LARGE HOSPITALS IN THE
UNITED KINGDOM.

BY

H. MACNAUGHTON JONES, M.D., M.CH.

Fellow of the Royal Colleges of Surgeons Ireland and Edinburgh;
Surgeon, Cork Ophthalmic and Aural Hospital;
Physician, Cork Fever Hospital, Cork Maternity, and Children's Hospital;
Late Senior Demonstrator (Lecturer on Surgical and Descriptive Anatomy) Queen's College, Cork;
President Elect, South of Ireland Branch of the British Medical Association.

LONDON:
H. K. LEWIS, 136 GOWER STREET.
DUBLIN: FANNIN & CO., GRAFTON STREET.
1876.

CONTENTS.

*THE fatal result which has lately so often followed the employ-
ment of an anæsthetic has prompted me to make these few
remarks on the use of anæsthetics generally, and the responsibility which the administration of such entails on the administrator. It is, I think, apparent to all, that the degree of
uncertainty which still appertains as regards the selection of
an anæsthetic, warrants its being made a subject of careful
consideration. It must ever be a matter of considerable
moment to an operator to select that anæsthetic which experience and experiment have proved to be the safest. If physiological experiment on the lower animals is borne out by
observation on the human subject of its effects, and that both
tend to prove the superiority of any one anæsthetic for the
great majority of surgical operations, it is certainly our duty to
avail ourselves of that one, when we desire to employ any. I am
the more desirous to bring this subject forward, as I am aware
that hitherto in this city (Cork) one anæsthetic, chloroform,
has been generally employed. The fact that from the introduction of chloroform into our city hospitals, to the present
date, not a single death has occurred from its use, speaks I
consider much for the care exhibited in its administration.
Nor, if we look merely to this lucky immunity from accident,
are we to be surprised, that those who have so long administered chloroform successfully, should hesitate before they
relinquish the use of this agent for any other. They naturally
are wedded, from a prolonged employment of it, and that a
satisfactory one, to this anæsthetic. But in addition, and this
it is which has chiefly induced me to make these few observations, there is an idea prevalent, that after all, there is a degree of danger about all anæsthetics, ·that one is only slightly
more or less dangerous than any other, that deaths occur ‚from
all, that medical opinions are divided as to the value and danger of each and all of them, and that even on their mode of

* Read before the City and County of Cork Medico-Chirurgical Association,
(Dec. 22, 1876).

action physiological observers are by no means unanimous. Again, there is the method of administration, and the agreeable nature of the anæsthetic. These points are naturally and rightly taken into consideration with the character of the operation to be performed under its influence. On these matters also various opinions are put before the profession, many of which tend rather to confuse than to settle the minds of men who are really anxious to arrive at a correct solution of the question. The only important issue in the matter is, what is the safest anæsthetic which should be employed, unless under exceptional circumstances, in all ordinary cases where it is necessary to administer an anæsthetic at all? It may appear perhaps foolish to dogmatise in such a case as the one before us. When evidence of the utility, and value of all anæsthetics in common use is so abundant; when the comparative harmlessness of each as contrasted with the deaths which would occur without its use is acknowledged, not alone by the profession, but by the public, and when many have for a quarter of a century used chloroform without a fatal result, it is not to be supposed that anyone anæsthetic would, even were its superiority indisputably proved, quickly supersede the use of all others. Nor, do I believe, will this ever be. The value of this or that anæsthetic, in many instances, depends it may be on the class of operation which is to be performed, the age of the patient, or other idiosyncrasy which may prompt the medical man to select a given one for a given case. These exceptional instances, however, only tend to establish a rule, that in all cases that it is to hand, and in which the operation can be performed successfully by its aid, the operator is bound to select that anæsthetic which is proved to entail the least risk to human life. This would appear so obvious a maxim that it is almost ridiculous to enunciate it here, were it not that I believe we have arrived at a certain knowledge of the comparative safety of these agents, and still do not act in accordance with it, but at times subject our patient to unnecessary and culpable risk. Our decision must I consider be based on the answer we give to these three questions, 1st, has it been proved by experience that any one anæsthetic, excluding all reference to rapidity of action, convenience, &c., is the safest? 2ndly, has this satisfactory conclusion been supported by direct physiological evidence, derived from experiments on the lower animals, and our knowledge of its action on the human economy? 3rdly, can it be availed of in the vast majority of cases, and administered with as great ease as any other anæsthetic, and with as favourable after results to the patient? If we reply in the affirmative to these queries, we must I consider hold ourselves responsible, if we do not give the patient the benefit of our convictions.

It is my intention to take up these three queries separately, and endeavour briefly to answer them as satisfactorily as our present knowledge of anæsthetic agents will permit me. I shall then refer to four agents for the relief of pain, viz., chloroform, ether, bichloride of methyline, and nitrous oxide. With the use of all four I have had personal experience, more particularly chloroform and methyline. I have but quite recently determined to use ether exclusively whenever it can be availed of. The nitrous oxide, I have done some few operations under, of a trifling nature, and cannot therefore speak from personal experience of its value. As this latter is chiefly employed in dental surgery, and as from the nature of its action it can never come into general use, it will not be necessary to do more than merely refer to it amongst the class of agents it is proposed to discuss. *Has it been proved by experience that any one anæsthetic is the safest?* I believe that it has, and I will briefly state a few of the recently recorded opinions on this subject. It is only within the last three or four years that the advantages of ether over chloroform or methyline have been forced on the notice of the profession in the United Kingdom, and that it has been adopted almost exclusively in some, and partially in many of the large hospitals. I was present in August 1872 when Dr. Joy Jeffries of Boston, in the United States, read a paper on the use of ether in Ophthalmic Surgery; I saw him on the same occasion administer it at the Royal London Ophthalmic Hospital, Moorfields. His method of administration, to quote his own words, was as follows, " a towel rolled into a cone, with a napkin or sponge pushed to the top of the inside is all we need to pour our ether on, whilst our fingers can mould it over any mouth and nose." "Some years ago I" (Dr. Jeffries), " heard in Europe medical men say ' But there are so many people who cannot take ether', I have yet to see one; the truth is I believe that surgeons who use chloroform are afraid of ether and do not dare to give enough of it in the commencement.

Now if the patient be warned that the ether will choke him, and told that when this occurs, to take long breaths to relieve it, and not to struggle and endeavour to push away the sponge, many will go to sleep quietly, and without trouble to themselves or the surgeon. I have but one other point to speak of, in reference to giving ether, when the patient, either old or young, struggles, and asks for a respite, and fresh air, do not yield. Hold him down by main force, if necessary, and at any rate keep the sponge tight over the mouth and nose, till he finally takes long breaths and then goes off into ether sleep. Doing this prevents him remembering anything about his struggles. It is absurd to stop the ether and try to reason with adults excited by the anæsthetic, and cruel not to push on quickly with children. This may seem puerile to my American brethren, but my personal experience tells me that those who use chloroform, may somehow have a dread of ether, as if it was to be suddenly fatal, and hence fail to give a patient enough to intoxicate him quickly. This arises from lack of familiarity with its use and administration."

I have thus quoted at length the remarks of Dr. Jeffries, as coming from a surgeon who had large experience of its use, and from the city "where its employment in surgery was discovered and promulgated," and also because I think he gives in these pointed comments all the most important hints and rules regarding its administration. As to its fatality "ether," he then stated, "is never fatal in surgery, that it is difficult to kill anyone with ether, and that death never occurs accidentally while it is being inhaled. The accumulated evidence on this point, is sufficient for me at least, as the accumulated evidence of the fatality of chloroform is sufficient to deter me from ever using it unless forced by necessity." Dr. Wood, in his work on *Therapeutics*, 1874, in the section devoted to anæsthetics, enters fully into the comparative safety of ether and chloroform. He says " so dangerous is chloroform and so safe is ether, that there is no excuse for the use of the former agent, under ordinary circumstances. The reason of the safety of ether, is that unlike chloroform it never suddenly paralyses the heart. It may kill by inducing asphyxia but it does so slowly and in the great majority of cases, after warnings, which can be overlooked only through the most reckless carelessness." Of chloroform he says, " It kills without warning so suddenly that no forethought or skill or care can guard against the fatal result.

It kills the robust, the weak, the well, the diseased alike, and the previous safe passage through one or more inhalations is no guarantee of its lethal action, statistics seem to indicate a mortality of about one in 3000 inhalations, and hundreds of unnecessary deaths have been produced by the extraordinary persistence in its use by a portion of the profession." He quotes Richardson in England, and Squibb in America in support of these statements. Looking nearer home we find that in nearly all the journals the use of ether instead of chloroform has been strenuously advocated. The *British Medical Journal* in particular first took the matter up some three years since, and has never ceased up to the present moment to urge on the profession the responsibility entailed in the administration of the latter agent. In an article on this subject in the journal I find the following : " Collecting from every source information as to the administration of ether as an anæsthetic we have invited from all quarters comment and communication calculated to completely inform the professional mind. The papers by Clover, Haward, Norton, Bowditch, and Fifield of Boston, Hutchinson, the late John Murray and a host of others, appeared to us to establish the superior safety of ether over chloroform, and they led to a very large and general substitution of ether for chloroform as a surgical anæsthetic." "We have never admitted that greater rapidity of effect, or other reasons of convenience could justify the use of a more rather than a less dangerous anæsthetic ; but inasmuch as the alleged inconvenience of delay, loss of time, and atmospheric diffusion of the vapour have undoubtedly produced a prejudice against ether, we think it to the point to observe that ether administered skilfully and properly is found in the hands of experienced operators, not to cause much, if any, greater delay than chloroform. Anæsthesia is we believe in practical hands produced at our hospitals by ether in about five minutes." " Surely" (says the same journal in commenting on one of the numerous deaths which have of late occurred from chloroform), " these incessant fatalities from the use of chloroform plead trumpet tongued for the adoption in its stead of what is vouched on high authority to be the safer anæsthetic." Dr. Lee of Chicago has stated that chloroform " is more generally employed than ether in the United States in consequence of its being more agreeable to the patient and rapid in its action ;

of 92·816 ether inhalations there were four deaths; of 152·260 chloroform there were 53 deaths, that is one in 23·204 of ether and one in 2·873 of chloroform, thus showing that chloroform is nearly 8 times as dangerous an anæsthetic as ether. Dr. Diday of Lyons asserts that any one using chloroform instead of ether is culpable. Dr. Morgan of Dublin, has more than any-one in Ireland advocated the use of ether in preference to chloroform. In a pamphlet published in 1872, he entered fully into the danger of chloroform and the safety of ether. He alludes to the report of the Royal Medical and Chirurgical Society on the therapeutical, physiological and toxical effects of chloroform, and he quotes the conclusion arrived at by that Committee, namely, that ether is less dangerous than chloroform and they admitted the greater danger of chloroform as compared with ether. He also quotes Hamilton of the Bellevue Hospital, New York, who in his treatise on *Military Surgery*, especially referring to the American war, bears testimony to the superiority of ether. He refers to its successful employment for ten years in the Bellevue Hospital, New York. Mr. Pollock of St. George's Hospital, where ether has got such an extensive trial, in a letter to the *Medical Press and Circular*, August 11th, 1875, says, "Ether has been administered at St. George's Hospital for the last four or five years in every variety of operation, and I can state without the slightest reservation that I have never personally witnessed in any single case an approach to a symptom indicative of anxiety." Professor Schiff of Florence, as we shall subsequently see on good grounds, "considers that with our present knowledge a practitioner is responsible for the death of a patient which occurs under ether; but is not so responsible when the death occurs during the inhalation of chloroform." He concludes from the physiological effects of chloroform that ether alone should be used, and chloroform rejected. On the continent, especially in France and Belgium it would appear that a conclusion has not as yet been generally arrived at as to the best anæsthetic. This was shown in the discussion which took place this year on the report by Dr. Duplay on M. Darin's *Memoirs on Anæsthetics.*

This report had chiefly reference to the protoxide of nitrogen, but several opinions were elicited as to the relative value of ether and chloroform. Dr. Perrin stated that ether was very little employed in Paris though he considered it would be better to follow the advice of the Lyons School and try ether, at the same time he believed the reason chloroform was continued in Paris was because it did so little harm.

Dr. Teulon, after an accident which occurred with chloroform, has for ten years employed ether, and administers it much in the same manner as that described by Dr. Jeffries, before quoted. Dr. Blot appeared to consider that there were great risks from large doses of ether, and that there are some cases in which it is not like chloroform, the quantity of the gas which kills, but in which there is present some idiosyncrasy of constitution in the patient, outside any affection of the heart, lungs, or nerves. Dr. Hélat thought that ether might produce accidents like those of chloroform. Dr. Giraldes, on the other hand, stated there were no accidents with ether at the same time that he confessed he had a weakness for chloroform. If we are to take the *Dublin Medical Press and Circular* as a fair criterion of the feelings of our Irish Metropolitan Surgeons on this point, they must be very strong indeed, as the annexed excerpt from this Journal will show. " The use of ether ought to be made a rule of practice instead of chloroform. We have seen and heard of too many deaths from the use of the latter anæsthetic to hesitate any longer in pronouncing that those persons who administer chloroform with the present knowledge of its dangers are culpably careless and ignorant." "The ether revival seems to have died out; not, we imagine, in consequence of the shortcoming of that agent, but because it is not found 'to pay' to take trouble about the saving of an occasional life." But it does not appear from the replies I have received from the various Dublin Hospitals that as yet the use of ether has entirely superseded that of chloroform. This will be apparent from the table appended to this paper. On the entire, however, it is manifest that in Dublin ether is receiving a fair trial, and that in many hospitals it is selected in preference to chloroform.

Let us now ask what those who have written specially on chloroform and advocated its use have stated on this important question of the relative danger of the two anæsthetics. The advantages claimed by Sir J. Simpson for chloroform over ether were briefly as follows : the quantity required being smaller ; its rapidity of action ; its pleasantness.; its cheapness ; its portability ; its ease of administration in the manner which he recommended, viz., "a little of the liquid diffused on the interior of a hollow-shaped sponge or pocket handkerchief, or a piece of linen or paper, and held over the mouth and nostrils so as to be fully inhaled." But even in midwifery practice Sir James Simpson apears to have recognized the value of ether as an anæsthetic, inasmuch as we find him in his essay "On the inhalation of sulphuric ether in the practice of midwifery," March, 1847, strongly advocating its use. As far back as that year we find him saying, that "abundant evi֊ dence has of late been adduced, and is daily accumulating in proof of the inhalation of sulphuric ether being capable in the generality of individuals of producing more or less perfect degree of insensibility to the pains of the most severe surgical operations." "I have employed it," he subsequently says, (November, 1847), "in every case of labour I have attended, with few and rare exceptions, and with the most delightful results." He here refers to the space of time which elapsed from his first introducing it to the notice of the profession until the writing of the paragraph quoted. "I have no doubt that in some years hence the practice will be general." "I have never had the pleasure of watching over a series of better and more rapid recoveries."

It is clear, that while not one of the advantages claimed by Sir James Simpson for the use of chloroform instead of ether, affects the all-important question of safety, which he judiciously avoided touching on, in the paper referred to, Sir James, himself, recognises the great value of ether and its use in the very department of medical science in which we might have supposed he would most strongly advocate the employment of chloroform.*

Dr. Sansom in his work on chloroform, 1865, says, "Ether is inconvenient, more than a pint of it is sometimes requisite to produce and sustain the insensibility necessary for an operation, whereas a few drachms of chloroform suffice." He alludes to the "excitement" and "sensation of choking," "the large quantities which escape becoming a nuisance to the operator and all around." He also refers to the fact that operations are commenced before the anæsthetic effect is completed under ether, from the large amount required; and also

---

* *Anæsthesia, &c.*, by Sir J. Simpson. *Article on Sulphuric Ether.*

he draws attention to the frequency with which chloroform has been administered as compared with ether, and that this point ought to be taken into account in estimating the relative fatality; at the same time he says that " ether being weaker is *a priori* a less dangerous substance than chloroform, but chloroform vapour freely diluted with atmospheric air can be rendered as innocuous as ether vapour." Dr. Sansom adduces the large number of cases in which chloroform had been administered all over the world up to the time of his writing, and the comparatively few deaths, as affording proof of the slight danger from chloroform inhalation. Thus he quotes Chapman who calculated that 1,200,000 operations had been performed in the ten years previous to 1859, when he wrote on the subject, and only 74 recorded deaths; the chances being thus 16,000 to 1, against a patient losing his life from the anæsthetic. He also refers to the French-Italian, and Crimean Wars, to prove the low mortality from the use of chloroform, and says that " the facts warrant the conclusion that the danger of chloroform is exaggerated." Still, Dr. Sansom's work is chiefly taken up with " the danger of chloroform," "diseased conditions which increase the danger," "danger of the incautious administration," "signs of danger," and "on resuscitation after death from chloroform," quite sufficient proof, if we required any, that Dr. Sansom fully recognises the danger which attends the administration of chloroform. Dr. Snow, in his work on anæsthetics, in writing of ether, says, "I believe that ether is altogether incapable of causing the sudden death by paralysis of the heart, which has caused the accidents which have happened during the administration of chloroform. I have not been able to kill an animal in that manner with ether, even when I made it boil, and administered the vapour almost pure. I hold it therefore to be almost impossible that a death from this agent can occur, in the hands of a medical man who is applying it with ordinary intelligence and attention." Mr. Clover, in concluding some remarks he made in the *Medical Times and Gazette* (June 27th, 1875), on the details of a case of death from chloroform administered for the removal of an adenoid growth from the posterior nares, says, "It will occur to many who have had experience of ether, that this would have been a safer agent, in a case where so much difficulty was expected; but it should be remembered that there was a special reason for avoiding ether, on account of its irritating effect upon the throat, even the small quantity of 4 minims in 1000 inches of air, seemed to increase the movements of swallowing when he first inhaled. Ether can indeed be introduced so gradually by the aid of nitrous oxide, as hardly to cause any irritation. I have so used it more than 400 times without any untoward result, and immensely less discomfort to the patient, than by the common method; however, I did not think this case fit for using any instrument or any combination that may be regarded as upon its trial. Perhaps it is fortunate that I did not, for it might have been more reasonably supposed to have caused death, than the methods employed (by Clover's apparatus), whose value has been established by satisfactory experience with them for many years. In the next case of this kind I shall give ether with nitrous gas, then, ether with a limited supply of air for at least ten minutes, and if I have any need to give chloroform at all it shall be even more diluted than it was in this case." Dr. Marion Sims in a valuable communication which he read at the meeting of the British Medical Association in 1874, on chloroform narcosis, having referred to the cause of death in many instances being cerebral anæmia, and the value of Nélaton's method of resuscitation by inversion, went on to say that "in America the accoucheurs often use chloroform, the surgeons mostly use ether." This he explained by the satisfactory results which follow the use of chloroform in labor, and the reason for which was to be found in the absence of cerebral anæmia during active labor when the opposite state, one of fulness of the cerebral vessels exists, no woman, he believed, had died in labor from chloroform. Dr. Sims thought "the safest plan is to relinquish the use of chloroform altogether except in obstetrics." The great number of deaths from the use of chloroform in surgical operations that have occurred even of late, should warn us to give up this dangerous agent, if we can find another that is as efficient and at the same time devoid of danger. "Ether fulfils the indications to a remarkable degree, but while it is safe, it is unfortunately unpleasant to the physician, and bystanders, as well as to the patient. He who will give us an anæsthetic as pleasant to take as chloroform and as safe as ether, will confer the greatest boon upon science and humanity." On the other hand I may remark that Professor Depaul has stated in the Académie de Médecine of Paris, that sudden deaths from the obstetric use of chloroform are not unknown to him. However this may be, they have not hitherto, I believe, taken place in this country. As the unpleasantness of ether has been made a prominent objection to its use, I desire to quote the practical remarks on this subject by Mr. Walter Rigden, formerly Resident Medical Officer to University College Hospital. In speaking of the most important differences in the after effects of ether, he says, "Vomiting after ether, I have found more frequent in my cases, but I am sure it is not so prolonged. Headache is decidedly more common after ether, and is sometimes intense but it is not usual. Some amount of excitement on returning consciousness after ether occurs in fully half the cases, but it is rare to be very great or prolonged." "There is no collapse after ether, and ether will in most cases do away with the shock after severe operations. Those patients who have taken ether subsequently to inhaling chloroform have told us that when consciousness returned they have felt much better after ether ; still the commencement of the inhalation of ether is so unpleasant that most say they would prefer chloroform if they were to be anæsthetised again. Vomiting I have found to occur in 50·65 per cent. of the ether cases, against 32·86 per cent. of the chloroform, but of these, 35·06 were only slightly sick, so that we have 15·59 per cent. of the ether cases exhibiting troublesome vomiting, against 16·17 of the chloroform cases." One fact worth noticing, is that alluded to by Dr. Richardson, that whereas in certain hospitals which formerly had only one death in 17,000 operations, there is now one in 1250. (Vide Dr. Balfour's letter to the British Medical Journal January 20th, 1872). "Chance or accident," Dr. Balfour writes, "is only another name for ignorance of the cause preceding or producing such an event," and he quotes Mr. Lister, who stated that "death from chloroform was almost invariably due to faulty administration." In the same letter he gives the details of a case in which death occurred from extraction of a tooth, though the patient had in all her confinements, eight or nine in number, taken chloroform. I do not fancy that many will be disposed to agree with Mr. Lister that carelessness does account for the numerous deaths which have of late taken place from chloroform. The history of all the fatal cases is directly against such a supposition, and it would be a most

dangerous one to permit pass uncontradicted. In a paper on
" death from chloroform" May 25th, 1872, Mr. Green, consult-
ing Surgeon of the Bristol Royal Infirmary, details the particu-
lars of seven cases of death from this anæsthetic and having
expressed his opinion of the value of galvanism in restoring
animation, he makes this observation "the use of chloroform
is a serious busines involving as it does, the issues of life and
death—how serious, few can realise, except those who have
seen one or more fatal cases." Mr. Erichsen in a clinical
lecture delivered at University College Hospital, June 1872,
stated that during the 25 years he was attached to the Hospital only one death occurred from chloroform. He divides
death from chloroform into three ·divisions. 1st. Death from
asphyxia. 2nd. Death from coma. 3rd. Death from syncope.
He believes that this latter mode of death is in reality one by
asphyxia, and that the heart is secondarily affected.

He thinks there is always an asphyxial tendency, but that
this is only marked in cases of feeble and fatty heart. He believes that the pulse, respiration, and color of the face must be
carefully watched. Dr. Sutton asks "how are we to know be-
forehand that a patient cannot without danger take chloroform? we have no physical signs that will guide us, and we
know that many persons with marked heart-disease will bear
the administration well." M. Bergeron gives the following
list, taken from the *Traite d'Anesthésie*, of Lallemand and
Perrin, of the causes of sudden deaths after ordering the use of
chloroform; 1, pre-existing organic affections; 2, nervous
state (emotion); 3, abuse of alcoholic drinks; 4, intervention of
surgery; to which I would add, as seems to be admitted, 5, a
peculiar influence of chloroform on the nervous organs, impossible to be foreseen, or to be remedied; 6, the use of chloral.
Thus it would seem that it is not possible to give chloroform to
anyone safely, because it can never be known who there is
who is not the subject of one or the other of these causes.
Hard drinkers, Dr. Sutton (and all other observers) remark,
bear chloroform badly and require greater care. In a paper
on the administration of ether, Mr. John Couper, assistant surgeon to Moorfields Hospital and Surgeon to the London Hos ·
pital, drew attention to the advantage of ether over chloroform
in various classes of operations, and particularly in eye operations. As at Moorfields twelve. operations are performed on

an average in the few hours each day, it becomes, as Mr.
Couper says, a matter of considerable importance that rapid
anæsthesia be produced. He refers to the speedy narcotism
and muscular quiescence resulting from ether, which, he says
are certainly as rapid as in the case of chloroform. " Thus
far" Mr. Couper affirms, " my impression is that ether is less
dangerous to life, on account of its inaptitude to produce car-
dial syncope; its administration is rarely if ever followed by
obstinate and dangerous nausea; it narcotises as quickly and
as certainly as chloroform; and with as complete muscular
quiescence; and that recovery from it is generally quiet and
satisfactory."* Dr. Jones disapproves of the combined use of
ether and chloroform : " one never knows what one is about."
He begins with chloroform, but as soon as the pulse shows
signs of failure, he commences the administration of ether, and
continues its use until the pulse is restored. He believes that
death from cardiac failure from chloroform may in this way be
averted.

Mr. Butlin chloroformist to the Hospital for sick children, in
a short article in the *British Medical Journal*, October, 1872,
confirms these opinions regarding the time requisite to produce anæsthesia with ether, and the absence of the subsequent
sickness present in chloroform. I consider the value of Mr.
Clover's method of administration of chloroform is best proved
by his own statement, that since for twelve years previously he
adopted the plan of measuring the chloroform, 3 minims and a
half to 100 inches of air, he had administered it without rejec-
ting a single case as unfit for it, and to many whose heart and
lungs were in a morbid state."† He never had a death from
it or from any anæsthetic.† Quite recently I performed seven
operations on the eye, within the hour, under ether; one of
these being enucleation, and another, the suction operation for
cataract. I have given ether, with Skinner's inhaler, several
times lately, and while I do not find it as rapid as bichloride of
methyline, nor in some respects, as easy of administration,
with the exception, in a few instances of some subsequent excitement, the after effects are perhaps better. I have had no
vomiting either during or after administration, and I have in
this period given ether to persons whose ages ranged from

* *British Medical Journal*, November, 1872.
† *Ibid.*, November 9th, 1872.

45 to an infant of 11 months; and strange to say, I found the latter patient the most difficult to anæsthetise. Dr. Nicholas Grattan has in a large number of cases administered anæsthetics for me for the past eight years, his long experience for the past 14 years both in the South Infirmary and County Hospital, and since he left this institution, at the Ophthalmic Hospital, gives considerable weight to his opinion. He is decidedly in favour of ether. I must confess that were it not for my firm belief in the greater safety of ether, I should from my experience be perfectly satisfied with bichloride of methyline in all eye operations, in which it was desirable and possible to give an anæsthetic. I have found that when the attention is divided between the effects of methyline and other anæsthetic, as must be the case where a large staff are not surrounding the operator, or a special anæsthetiser, there is a serious drawback to the quickness, and, it may be, success of the operation. This is still more the case if the patient be a bad subject (some old inebriate), for its exhibition.

To any one then situated as I often am, in operating without many assistants, the boon of a safe and reliable anæsthetic must be a considerable one. I have found almost without exception, that children bear either chloroform or methyline admirably. I have never had in a child a case in which I was even alarmed. I now come to other anæsthetics which are employed by surgeons in various operations and have been used instead of chloroform or pure ether. I cannot enter fully into many matters of interest with regard to these but will briefly summarise the conclusions which have been arrived at by leading anæsthetisers regarding them. I have no personal experience of a mixture of ether and chloroform, or chloroform and rectified spirit, and am not aware of the number of fatal cases which have been recorded. That death has taken place from this method of administration is certain, and as far back as July, 1857, a death occurred in America from chloroform and ether, but which Mr. Snow has shown really happened in consequence of hæmorrhage, and not from the anæsthetic. He also disapproves of this method in consequence of the different degrees of density of the two vapours, so that in the commencement ether chiefly is inhaled and secondly chloroform. The practice of giving a few whiffs of chloroform in the first instance, and then administering ether has found favor with many; I have myself so given it with the best result. The two agents which have been principally employed to produce rapid anæsthesia, within the past few years, are bichloride of methyline and nitrous oxide gas. The former introduced by Dr. B. W. Richardson in 1867, has been extensively employed, and not alone in short operations, but also in such operations as that of ovariotomy. Of this agent I have had myself considerable experience. I have been using it constantly for all short operations in hospital and private practice for over 7 years. I cannot have given it less than between 1200 and 1300 times.

I have already in the *Dublin Medical Press and Circular* recorded my opinion of its use and advantages over chloroform for short operations. Though I have used it chiefly for eye operations, I have also performed a variety of other operations by means of this anæsthetic. I have had no fatal result, I have had cases, in nearly all instances, hard drinkers or old tipplers, who bore the anæsthetic badly, and on some occasions I have been alarmed and have had to desist from the administration. In any old cases of chest affection or those with a history of past bronchitis and asthma I have also found the administration dangerous and had to discontinue it. But with these exceptions I must say I have had the very best results from the use of methyline. I have found it rapid, not so unpleasant to the patient, produce perfect anæsthesia, and decidedly in my experience the after effects are not at all so unfavourable as those of chloroform. I gave it at first in a long flannel bag fixed with wire. I have of late years administered it in the conical gauge bag lined with flannel containing a small sponge, made for me by Messrs. Mayer and Meltzer. I may say here that I have found no method of administering *chloroform* yield greater satisfaction than that with this simple bag. I have given it many times with it and have been always pleased with the result. Snow's inhaler I have nearly always used in obstetric cases and also in many operations which required prolonged anæsthesia. This inhaler I have also found act admirably and it has the great advantage over the lint plan of preventing the escape of chloroform and saving the vapour. I believe that there is not one of Clover's apparatus in this city, and that the method of administration adopted has been principally that suggested by Sir James Simpson, by means of the handkerchief, or lint. I have given chloroform several times in this way, but it has many disadvantages, as for example, the diffusion of the chloroform in the surrounding atmosphere, the difficulty of getting the patient to inhale, the length of time it is necessary to prolong the administration, the uncertainty of the dilution of the chloroform vapour, which varies greatly at times, the waste of the anæsthetic, the struggling and excitement of the patient, which in my experience are greater with this method of administration than with either of the methods before mentioned.

Bichloride of methyline has been largely used elsewhere, and as I have already stated in protracted operations. I have seen Mr. Spencer Wells successfully perform ovariotomy by the aid of this anæsthetic at the Samaritan Home. On that occasion, though it was necessary to keep up the influence of the anæsthetic for some time, it answered admirably. Mr. Lawson Tait has also operated with bichloride of methyline many times for ovarian disease, and he has also used the methylic ether, and speaks highly of the suitability, of the latter anæsthetic for this operation. Mr. Richardson claims these advantages for methyline over nitrous oxide; it is not necessary to exclude completely atmospheric air; it is not a purely asphyxiating agent which must if prolonged beyond a certain period, necessarily kill; it does not require a costly and troublesome apparatus. Respecting bichloride, Dr. Richardson has stated that he was "not favourably impressed with its application when very quick anæsthesia was required. That it was rapid in its action was true, that it answered the end it had in view was true, and that it had now been used, for rapid inhalation, an immense number of times was also true; but these facts could not conceal the further and all important fact, that the bichloride of methyline belonged to a dangerous family of chemical substances, and could not therefore be played with without risk. It had been extolled as safer than chloroform, and that was allowed, for as it contained one equivalent of chlorine less than chloroform it was materially safer, but the safety was relative, not absolute." The value of the latter remark is enhanced when we consider the fact that commercial chloroform contains as a rule compounds of chlorine, and according to some authorities* pure chloral hydrate, is the only substance from which we can get pure chloroform. The purity of the chloroform and its resulting decomposition, if retained, are matters worth the serious attention of those who still continue to use chloroform on all occasions when they require an anæsthetic. Dr. Richardson, struck with the disadvantages attending the use of bichloride of methyline, experimented on a variety of substances with a view to ascertain the safest, and at the same time most rapid anæsthetic, and in the year 1870, he brought before the Medical Society of London the results of his researches. He decided in favour of methylic ether. He inhaled it himself, was "narcotised completely in one minute, was unconscious in seventy seconds, and recovered almost instantaneously without nausea, headache, or other unpleasant symptom." In the extraction of teeth, from one to three minutes suffice from the commencement of the inhalation to the termination of the operation, and there have been cases in which the entire process has been gone through in 45 seconds. It has all, Dr. Richardson asserts, the advantages of bichloride of methyline over nitrous oxide, and is superior to the former as it does not produce muscular spasm or syncope. For some years, bichloride of methyline has received the fullest trial at Moorfields Hospital. They now use ether at this institution. The large number of operations annually performed at this Hospital, all of which require rapid and complete anæsthesia, compared with the few deaths which have occurred in it, from methyline, afford the best test of its safety and value. But it is clear from the fact that this anæsthetic has been relinquished for ether, that they do not consider it as safe as the latter. Within the past two years, two deaths occurred at Moorfields from this anæsthetic. In neither case was there any indication of danger from the state of the pulse or heart. In the last instance death occurred from the exhibition to a healthy sailor aged 27, of one drachm and a half of the methyline.

Mr. Jabez Hogg, who has used bichloride of methyline extensively in eye operations, has reported most favourably of it, though he stated at the Congress of 1872, that he has nearly relinquished their use, preferring to operate, in a large number of cases without them. I must say that of late, I follow the same rule, and in many cases operate without anæsthetic.

* Dr. Hager, *Year-book of Pharmacy*, 1870.

C

This may not be so, when I have a larger experience of ether. Mr. Hogg stated some time since that out of 2000 cases of methyline administration he had not one bad result. I am not aware of the exact number of deaths recorded from bichloride of methyline. Up to 1874, four deaths were recorded,[a] and since then, the two deaths above referred to, have taken place at Moorfields. But this is a very small mortality if we take into consideration the several thousands of times it has been employed during these eight years. As regards nitrous oxide gas, it is generally admitted that as Dr. Wood says, "its administration is simply a neat way of stopping the supply of oxygen, and thereby inducing asphyxia." It may have some other less important effects on the nervous system as the primary excitement tends to prove, but that its action as an anæsthetic is due to its simple asphyxiating influence, no one doubts. "Nitrous oxide," Dr. Wood says (1874), "has been administered to many thousand persons, and until recently no deaths at all attributable to it have occurred." The cases which have been reported, he doubts (with one exception), if they can be really attributable to the gas. Of 300,000 cases published by the States, not a single case of death occurred; at least, so M. Darin stated at the discussion. The short period of the anæsthesia and the nature of the apparatus, M. Darin states, have caused it to be abandoned. On the other hand, it has been asserted (same discussion) by Dr. Duplay, that Marion Sims published a case of ovariotomy, which lasted one hour and a half under nitrous oxide. However this may be, any one who has ever seen nitrous oxide administered, and watched its effects, must be struck by the asphyxiated look which the patient presents. So terrible is the appearance, so horrible is the death-like lividity of the features, that it is almost impossible for a friend or relative to watch the instantaneous change which spreads over the face without great alarm and shock. I have seen it several times administered with great care and skill by Mr. Corbett, jun., of this city, and I have done some trifling operations by the aid of this anæsthetic administered by this gentleman. One thing struck me in the cases I have seen, and that is, that of all anæsthetics I have

* *Medical Times and Gazette*, 1869, vol. ii., p. 524. *Brit. Med. Journal*, Sept. 1871, August 1872 and Oct. 1872. Also Wood. Chapter on Anæsthetics.

seen used, it is the one in which the greatest tact is required both to prevent the alarm of bystanders, and to secure the instant performance of the operation when its effects are matured. The committee appointed in 1872 to enquire into the action of nitrous oxide, by the Odontological Society, got into their possession the records of 58,000 cases in which the gas was administered in this country, and they stated that not a single fatal case had occurred which could be fairly attributed to its use. The committee were of opinion, that in operations on other parts of the body than the mouth, by checking and resupplying the gas through the face piece from time to time, as circumstances may require, insensibility can be kept up for several minutes; they expressed their conviction, that pure nitrous oxide properly administered is "the safest anæsthetic known."* In writing on the same subject, Dr. Jones, formerly administrator of anæsthetics at St. George's Hospital says, "when I consider the safety, ready applicability, and the pleasantness of this agent, together with the success attending it, I confess I am surprised that Surgeons do not adopt it more generally." "I have succeeded in keeping cases under its influence for eight or ten minutes by its reapplication from time to time."†

In reference to ovariotomy, I may draw attention to the table of cases published in the *British Medical Journal*, June 26th, 1875. In 50 cases, from October 1872 to May 1875, there were 6 deaths. All the operations were performed under sulphuric ether. I have ever found women more tolerant of all forms of anæsthetics than men. I am not in a position to say positively, but I am under the impression that deaths have occured a far greater number of times in males than in females. I before alluded to Dr. Marion Sims' reference to the tolerance which there is for chloroform during the obstetric effort. Dr. James Campbell, late House Surgeon to the Maternity, Paris, and "Chef de Clinique Obstetricale" in the Paris Faculty, has written an interesting article on the subject in the *Practitioner* of May, 1874. He shows, quoting Professor Béclard, that there exists, during the efforts of expulsion (when chloroform should be administered), a decided hyperæsthesia of the brain and other organs, and that there is established an antagonism be-

---

* *British Med. Journal*, November 9th, 1872.
† *Ibid.*, November 30th, 1872.

twixt this condition and the anæsthetic anæmia in which there
are fluxes and refluxes of blood to and from the brain during
the contractions, and between the intervals of the convulsive
efforts. He concludes that there is a peculiar anæsthetic
tolerance, which parturient women seem to enjoy in all cases
under the care of an accoucheur, within the experience of the
last quarter of a century, to the dangers of anæsthesia." Dr.
Wood in his allusion to bichloride of methyline, says, that it
has never been employed in America, and appears to think
that its use as an anæsthetic is limited in greater part to
London. Dr. Richardson, he says, "with rare control expressed no opinion as to the real safety of this new anæsthetic,
and subsequent events have justified him." I have I think
shown sufficiently that if Dr. Richardson could have foreseen
subsequent events, they would have justified him in speaking
confidently of the comparative safety of methyline and chloroform and the success which attended his discovery. But Dr.
Richardson had sufficient control not to introduce bichloride of
methyline to the profession until he had himself frequently
inhaled it, and had repeatedly subjected the lower animals to its
influence with perfect safety. But in addition the concluding
remarks of Dr. Richardson (November 1867) show that he had
carefully calculated the relative safety of this and other anæsthetics. "Its safety as a general anæsthetic must therefore be ac-
cepted as relative rather than absolute; I have tried to ascertain
its relative value with as much care and candour as I could sum-
mon, and the result of my work leads me to hope that the balance
of safety is on the side of the bichloride". Three observations
bring me to this reasoning; 1st, I find that if two animals of
the same kind and age, say pigeons, be placed in chambers of
the same size and exposed to the same temperature, and under
conditions the same, to equal value of chloroform, tetiachloride
of carbon, and bichloride of methyline, the resistance to death
will be as 14 to 5 in favour of bichloride of methyline,
against the tetiachloride of carbon, and as 14 to 9 against the
chloroform. In 1871 Dr. Richardson introduced hydramyl as
as an anæsthetic for short operations. He experienced on
himself and gave the hydramyl to others. Mr. Mathews extracted a tooth under its influence, and the whole proceeding
from the administration of the anæsthetic until return of consciousness did not occupy more than a minute; the tooth was

extracted at the expiration of thirty seconds. Subsequently on account of its great volatility Dr. Richardson added chlorine, and the new anæsthetic hydramylchlor produced insensibility in 40 seconds. It appeared to have this advantage over nitrous oxide, in addition to the physiological difference, that the sleep was perfect and the face underwent no change whatever. Reviewing now the evidence I have adduced, and which bears directly on my first question, I consider we are in a position fairly to answer it in the affirmative. Ether has undoubtedly been shown, and appears now to be universally acknowledged to be the safest, anæsthetic. If then we desire rapidity of anæsthesia for special purposes, such as dental operations, nitrous oxide appears from past experience to be equally safe, though its mode of action being unfavourable, such an anæsthetic as methylic ether may yet be preferred. In obstetric practice as there appears to be a special tolerance for anæsthetics, either chloroform or ether may be given, as I am not cognisant of any authentically recorded death from the former, during its administration in in the obstetric condition.

Have these satisfactory conclusions been supported by direct physiological evidence, derived from experiments on the lower animals, and our knowledge of its action on the human economy? I shall summarise the most important effects which have been ascertained in reference to the three anæsthetics above named. (These conclusions are epitomised from Wood's work on Therapeutics). It has been proved by experiments on the lower animals that chloroform produces cerebral anæmia and that the muscular excitement and couvulsions are cerebral.* It first produces diminished action of the heart, by stimulating the inhibitory centres, and secondly rapidity of action by paralysing them.† It produces decided lowering in the arterial pressure.‡ It appears to act on the vaso-motor nerves, as on the vagi, first stimulating, secondly paralysing. It exercises a directly paralysing effect on the heart muscle.* Chloroform produces an integral change in the blood cells; this change may be chiefly one of mere contraction due to oxidization.§

Ether. On the lower animals ether acts precisely as it does

* Carter, Bernstein, Bert. (H.) Wood.
† Glover, Gosselin, Anstie.
‡ English Chloroform Committee.
§ Harley, A. Schmidt, Schweiger-Seidel and others.

on man, (Wood). From the experiments of Flourens and
Longet it has been conclusively established that ether acts first
on the cerebrum, secondly on the sensory centres of the cord,
3rdly the motor centres of the cord, 4thly the sensory centres of
the medulla oblongata, 5thly the motor centres of the medulla
oblongata. Ether increases the arterial pressure. Ether in
all probability stimulates the vaso-motor system and increases
the power of the heart.

Nitrous oxide. The experiments of French observers, M.M.
Jolyetd and T. Blanche on various animals tend to prove that
this gas acts by depriving the blood of oxygen. This has
been also proved by an analysis of the blood of dogs rendered unconscious with the gas, the oxygen being consider-
ably diminished, whereas the amount of carbonic acid was
not increased. All observations tend to prove that the asphyxiating property is the essential one in nitrous oxide.

If we look impartially on these physiological effects, we
find ample cause for death from chloroform and safety
from ether. In the one, the cerebral anæmia, the paralysis of the inhibitory centres, the lowering of the arterial
pressure, the secondary paralysis of the vaso motor nerves;
in the other, the motor centres of cord and medulla being
last acted on, the increase of the arterial pressure, the sti-
mulation of the vaso motor system, and the increased power
of the heart. It will ever be, as it has I believe ever been, a
matter of speculation, and one dependant on the peculiar
balance of power existing in certain individuals anæsthetised,
between the heart and nervous centres, whether death takes
place from chloroform or not. And not alone between the
heart and nervous centres, but on the integrity of all the tissues,
of the heart tissue, and that of the vessels, and of the blood
itself. The cause may not lie in an unhealthy state of any
one of these, but in some condition of abnormal relationship
which any two or all of them bear to each other. Practical
experience teaches us the truth of this, as neither by physical
appearances or symptoms are we able to prognosticate the
case in which chloroform is likely to produce a fatal issue.
These states are beyond our ken, and hence the great danger of
the anæsthetic. In an important lecture on " one of the causes
of death during the extraction of teeth under chloroform", Dr.
Lauder Brunton has recently pointed out, how a little chloro-

form administered in the sitting posture may be very dangerous to the patient, who is at the same time submitted to a severe shock to the large sensory nerve, the fifth. In fact how the administration of a little chloroform at all, is a practice to be condemned. He shows how this small quantity of chloroform acts first on the cerebral hemispheres, and relaxes the vessels while it permits the reflex stoppage to go on in the heart. In the waking state, the contracted condition of the vessels requisite for the due supply of blood in the veins is maintained ; under chloroform it is lost, and if while in this state, the shock is transmitted by the evulsion of a nail or the drawing of a tooth, the heart will stop in consequence of the combined causes, and death ensue. He thus draws attention to Professor Syme's maxim : "always use good chloroform and give enough of it." It is needless to say that these remarks of Dr. Brunton do not affect the general question of the safety of chloroform, but duly tend to point out a common source of danger, viz , the administration of small quantities of chloroform in minor operations. I may here draw attention to the valuable communication of Dr. Hake in the *Practitioner*, of April 1874, on the studies of ether and chloroform from Professor Schiff's Physiological laboratory. These observations of Professor Schiff were conducted in the most complete manner for a number of years, and they leave no doubt as to the answer we must give to my second question. They confirm completely the summary I have given above of the relative physiological effects of chloroform and ether.

Dr. Schiff's conclusions have been chiefly derived from the physiological experiments he has been conducting for various purposes in his laboratory at Florence and for the correct carrying out of which it was requisite to give an anæsthetic; I shall only make three short quotations from this paper ; "we adopt ether and not chloroform," he says, " because a very extensive experience has shown that etherization pushed to the very last stage of insensibility is never dangerous to life, as long as one maintains the act of respiration," and even if pushed beyond this point, he points out how artificial respiration is sufficient to prevent life being extinguished.

"We are able to say in the present state of science that a medical man is responsible for every case of death occasioned by the application of ether, because a careful watching of the respiration is capable of preventing death, whilst the effect of chloroform depends in part on individual predisposition which the physician is unable to recognize. Our own experiments bearing on this argument enable us to say that in more than 3000 cases we have adopted ether with a view to preserve the life of animals, and with the few exceptions indicated elsewhere, not a single case of death occurred. On the other hand chloroform has cost us a considerable number of animals when I have wished to push anæsthesia to its utmost stage." I say then that we may fairly answer the second question in the affirmative.

3rd. Can ether be availed of in the vast majority of cases, and administered with as great ease to the patient and with as favourable after results ? It is not now necessary to enter at any length into an answer to this question. I believe I have in the foregoing remarks proved and will show by the list of the hospitals which now use ether alone, or some combination of it, and as we may also infer from its extensive use in America, that it can be availed of generally. There may be some difference of opinion as to the comparative pleasantness of this agent and chloroform but there is none as regards its superiority in the after effects. I believe then that I have shown that we must also answer question number three in the affirmative. Having now entered, I regret to say, but very inefficiently, into the debated question of the safety of chloroform and ether, as proved by practical experience in the human subject, and experiments on the lower animals, and the adaptability of the latter to general surgery, I propose to make a few concluding remarks on the responsibility of medical men in the employment of anæsthetics. I have avoided entering fully, in this paper, into the various methods adopted for administering either chloroform or ether. This much I must say that I consider that any one who continues to administer chloroform, especially in a large hospital should be provided with a Clover's apparatus, as by it alone can be measured the exact proportion of the chloroform to the atmospheric air. If I am told that nothing is safer and so simple as the folded piece of lint or handkerchief, I would simply refer to the deaths which have over and over again occurred from this mode of giving it, and to the list of 50 fatal cases published by Dr. Snow in which by far the majority of deaths will be found to have occurred when the chloroform was administered in this method. I can only say that to me it is wasteful, clumsy and uncertain, and in the very nature of the method, from the unequal diffusion and exhibition of the vapour, dangerous.

This may not appear so to others and I do not wish to press the point. I much prefer the gauze inhaler. For obstetric cases I can conceive nothing superior to Snow's inhaler. I may here notice the proposal of M. Forné*, to give hydrate of chloral previously to the exhibition of chloroform, a position which met with strenuous opposition from Messrs. Dolbeau, Demarquay, Séé, and others, while eminent surgeons, such as Nussbaum, Rigault, and others, have successfully applied it. The idea originated with Claude Bernard. It does not appear to have been received with favour, and certainly on physiological grounds, considering the cause of death from chloroform it would appear to be the agent, of all others, contraindicated. I have shewn Skinner's ether inhaler. There is also the inhaler invented by Mr. Morgan, and another by Mr. Richardson of Dublin. There is also the inhaler sold by Hawkesley and which is called after his name. Some prefer the American system as described by Dr. Jeffries. Whatever one is adopted, the secret (as Mr. Morgan says), is to exclude air, and give plenty of ether.

We must perceive that we may narrow the question of responsibility in the administration of anæsthetics and divide it into two heads. 1st. The kind of anæsthetic employed. 2nd. Its mode of administration. I have chiefly to say to the former. The administrator of chloroform can certainly fall back on the assertion, that no matter in what form it is exhibited, it may produce death, and that it has done so, when the greatest of living surgeons, and also many who have passed away, were superintending its use. On the other hand, he who gives ether has to accept a certain amount of responsibility, as up to the present, all the weight of evidence is against *him*, and in favour of the anæsthetic in the event of a fatal issue. It certainly will ever be a matter of importance for the operator to prove: 1st, that the most approved method of administration (where possible), was the one adopted: 2nd, *that every known or available means was resorted to, to resuscitate the patient*

before life was despaired of. But a much more serious consideration for medical men in future will be, what excuse they
will offer for not employing the anæsthetic which has been
proved to be the safest under all circumstances for the relief of
pain. Will the convenience or prejudice of the operator, or a
yielding to the feelings of the patients, or the effort to make
anæsthetism a luxury, save the administrator or operator from
an imputation of being culpable, and one which may seriously
damage his professional reputation ? I am afraid not.

True, he may prove all that I above stated in reference to
deaths from chloroform, but this fart will only serve the more
to fix his responsibility in administering it. We may know that
thousands of patients have succumbed from shock, who never
got any anæsthetic, and many more from shock and not from
the anæsthetic given. But this is no argument when the question is one of responsibility for the use of the safest anæsthetic.
"An American Jury," says Dr. W. C. Fifield, of Boston, in
alluding to the responsibility entailed in the administration of
chloroform "would probably consider the matter in this light,
viz., that whosoever administered chloroform by inhalation for
relief from pain, knowing that sudden death may result from
causes too numerous to mention, and even when no cause can
be found for such death other than the agent employed, knowing also that another agent equally capable of producing such
anæsthesia exists, which has been shown to be far more safe
than chloroform, viz., sulphuric ether, shall be held 'criminally'
responsible for his temerity." ⁕

I cannot conclude this paper better than by quoting his (Dr.
Fifield's), forcible remarks on this subject.

"Do not all this paraphernalia of air-bags and special apparatus, these proposals to give chloral and opium before inhalation, in order to lessen the amount of chloroform employed,
this timid one-fourth and one-half anæsthesia, bespeak the in-
ward conviction, the instinctive knowledge, that death from its
use may come, and come quickly and immediately, and none
can tell why or whence it came ? Let us, then, see no more
tears, or hear no more cases of fathers and mothers, of

⁕ I strongly object to the introduction of the word " *criminally*." *Hitherto* the question of the employment of one or the other anæsthetic in various
cases has been too much *sub-judice*, that any man should be held even *culpable* for the use of any one in particular.

widows and orphans, who weep their dead; the dead who, but a moment ago, lay down trustingly for the reduction of a dislocation, the opening of a whitlow, or the removal of a wen. Let us employ ether, to which no reproach exists, except its evil smell. Unpleasant it may be, but its unpleasant *smell* is not to be compared to the unpleasant *sight* of a corpse; a corpse that, but for the employment of chloroform, might still have been a living man or woman, or child. It has been objected that ether costs more. The expense of one funeral would pay for a deal of ether."

The responsibility, with our present knowledge, from henceforth is greater. Medical men who continue to use chloroform *alone* must not forget this.

In very young patients the good result of chloroform may tempt many surgeons to employ this agent, and certainly past experience shews but a very remote risk in the instance of children; and in young patients the combination of ether and chloroform, is more to be urged, as the addition of chloroform and its rapidity of action do away with much of the terror of the operation.

REPORT OF KIND OF ANÆSTHETIC USED IN SEVERAL LARGE HOSPITALS IN THE UNITED KINGDOM, AND OBSERVATIONS.

Name of Hospital.	Anæsthetic Used.	Remarks.	House Surgeon.
Adelaide Hospital, Dublin	Ether by means of Richardsons's apparatus; also chloroform administered on lint.	Both satisfactory.	H. J. Battersby.
Richmond Hospital, Dublin	Chloroform. Ether lately introduced·	Not capable as yet of forming an opinion on ether.	Not given
Sir P. Dunn's Hospital, Dublin	Chloroform.	Most satisfactory. Skinner's inhaler and a dropping bottle used.	J. Barton.
Jervis Street Hospital, Dublin	Chloroform	Perfect in its results, administered with a plain wire framed inhaler, with free admission of air.	R. D. Kenny.
St. Mark's Ophthalmic Hospital, Dublin	Chloroform.	Satisfactory.	J. O. B. Williams.
Stevens' Hospital, Dublin	1. Chloroform. 2. Ether principally used. 3. Mixture of chloroform and rect. spirit.	1. Most satisfactory, specially with children. (2) Generally satisfactory, but the patients troublesome. Data not sufficient to gixe opinion.	Henry J. Tweedy, *Surgeon.*
National Eye and Ear Infirmary, Dublin	Ether generally, chloroform sometimes.	Satisfactory, except for length of time in putting the patient under its influence, and sick stomach ensuing. Richardson's apparatus used, requires too much ether and does not sufficiently exclude the air. Anæsthetics not much used. No anæsthetic for cataract or iridectomy; only used for such operations as enucleation, plastic, or painful or protracted ones.	H. R. Swanzy, *Surgeon.*
Royal Sick Children's Hospital, Edinburgh	Chloroform.	Satisfactory, given on a conically folded towel.	Jas. Crabb.

Name of Hospital.	Anæsthetics Used.	Remarks.	House Surgeon.
Glasgow Royal Infirmary	Chloroform.	Satisfactory.	M. Thomas, *Superintendent.*
St. Vincent's Hospital, Dublin	Ether. Chloroform administered only in exceptional cases.	Properly administered, quick, and satisfactory; if not so, tedious and sometimes fails altogether. A large quantity of the vapour given at once.	F. B. Quinlan, *Surgeon.*
Mater Misericordia, Dublin	Ether.	Satisfactory; several inhalers, sub-judice.	T. Staunton.
Aberdeen Hospital	Chloroform.	" Fairly" giving satisfaction.	J. Smith.
Belfast Royal Hospital	Chloroform.	Given over 5000 times, not one death. Administered on a sponge.	Stanley Coates.

First Hospital in Ireland in which it was administered. In constant use.

Name of Hospital.	Anæsthetics Used.	Remarks.	House Surgeon.
Mercer's Hospital, Dublin	Chloroform.	Satisfactory.	M. H. Molahan.
City of Dublin Hospital	Anhydrous ether	Partially satisfactory, objection to slowness of action and sick stomach which ensues. Richardsson' apparatus used.	G. K. D. Charlton.
Meath Hospital. Dublin.	Chloroform, sometimes ether.	Both satisfactory.	J. Atkinson.
Birmingham General Hospital	Ether, with adults principally, and chloroform with children.	Both satisfactory. Ether on an inhaler. Chloroform on lint.	Walter Ottley.
The Queen's Hospital, Birmingham	Ether.	Satisfactory, administered by means of a folded towel formed into a cone, with sponge.	T. MacQueen
Adderbrook's Hospital, Cambridge	Ether.	Safe but unpleasant.	G. E. Wherry.
Royal Southern Hospital, Liverpool.	Chloroform, ether occasionally.	Both satisfactory.	Henry Caddy.
Royal Infirmary, Manchester	Generally chloroform, occasionally mixture of sulph. ether, chloroform and rect. spirit.	Perfect satisfaction. Inhaler not used. (Reply torn).	Charles Edwin
Metropolitan Free Hospital, London	Chloroform.	Satisfactory. Given on lint.	F. D. Hayman.

NAME OF HOSPITAL.	ANÆSTHETIC USED.	REMARKS.	HOUSE SURGEON.
Sick Children's Hospital, Gt. Ormond St., London	Chloroform.	Satisfactory, administered on lint.	W. E. Cant.
"London" Hospital	Ether. Chloroform with children and old people.	Satisfactory. Patient made to inhale a large quantity rapidly.	J. H. Allden.
University College Hospital	Combination of nitrous oxide gas and ether.	Most satisfactory. Clover's gas and ether apparatus used.	Walter B. Houghton.
Westminster Hospital	Chloroform and ether, alone, or in combination.	Satisfactory. Clover's apparatus used for chloroform, and an inhaler of sponge, cotton wool, and oiled silk for the ether.	Arthur Price.
Bristol Royal Infirmary	Ether.	Satisfactory, a large sponge medium used.	H. M. Chute.
Bristol General Hospital	Chloroform.	Satisfactory. Plated inhaler.	W. H. Harsant.
Leeds General Infirmary	Chloroform and ether, the latter more frequently.	Objection to ether, length of time taken to administer it, and objectionable in case of pulmonary disease or tendency thereto, and action not so deep as that of chloroform. Chloroform given on a piece of calico or lint, ether on a conically folded towel or by Dr. Allis' inhaler which is very efficient.	George Powel.
Newcastle on Tyne Hospital	Chloroform chiefly. Ether lately commenced.	Chloroform satisfactory.	G. F. Beatson.
The Evelina Hospital, London	Chloroform (entirely).	Satisfactory.	W. E. Paley.
St. George's Hospital, London	Ether.	Satisfactory administered by means of a felt cone with sponge at top.	W. H. Bull.
Royal London Ophthalmic Hospital	Ether.	Satisfactory as a rule.	A. Stanford Morton.
Middlesex Hospital	Pure ether.	Entirely satisfactory, administered with Hawksley's apparatus.	A. B. Scott.

Name of Hospital.	Anæsthetic Used.	Remarks.	House Surgeon.
St. Bartholomew's Hospital	Ether, preceded by nitrous oxide gas.	Satisfactory. Only one case of sickness, (lasting 12 hours) out of 800 or 900 cases). Administered with Clover's gas and ether apparatus.	Joseph Mills.
Hospital for Diseases of the Throat, Golden Sq., London	Chloroform up to the present, ether intended to be introduced.	Skin frozen in Tracheotomy.	W. Pugin Thornton. *Surgeon.*
Guy's Hospital	For children chloroform, invariably, and for adults frequently, sometimes the mixture, alcohol 1 part, chloroform 2 pts., ether 3 pts., when surgeon prefers it, or chloroform badly taken. Ether in hazardous cases.	Chloroform most satisfactory. 'The pupils useless as a guide to change but the conjunctiva useful to show when patient is unconscious. Good plan to place patient under the chloroform and keep him under the influence of the mixture.	J. Farrant Fry.
King's College	Chloroform and ether equally.	Satisfactory.	Not given.
St Mark's Hospital, London	Ether.	Entire satisfaction.	W. J. Roeckel.
Samaritan Free Free Hospital	Bichloride of methyline.	Very satisfactory. Dr. Junker's apparatus.	G. Scudamore, *Secretary.*
Dublin Eye and Ear Infirmary	Ether exclusively.	Perfect satisfaction if properly administered. Air must be altogether excluded if possible. Morgan's inhaler alone effects this purpose. The ether must be the best and the patient's stomach empty.	A. H. Jacob. *Surgeon.*
Edinburgh Royal Infirmary	Chloroform.	Satisfactory.	T. Spence. *(For the other House Surgeons).*

See Addenda for other Hospital Reports.

ADDENDA.

Since I read this paper and received the reports here published, the *British Medical Journal* has instituted a series of enquiries on the following points.

British Medical Journal, Dec. 25th and Jan. 1st.

1. What anæsthetics are now in use, and for what cases is either anæsthetic preferred?

2. What methods of administration are employed?

3. Has any change been made in the last four or five years in the anæsthetic used, or its mode of administration, and if so what were the reasons for the change?

4. Can any suggestions be made by the adoption of which the safety of the anæsthetised person might be more completely secured, or any improvements in the production of anæsthesia for surgical operations be effected?

The following replies have up to the present been received.

St. Bartholomew's Hospital.—Mr. Joseph Mills (Chloroformist to the Hospital) reports as follows. The anæsthetics now in use at St. Bartholomew's Hospital) are ether, chloroform, and nitrous oxide gas; ether preceded by nitrous oxide gas is used in by far the greater nnmber of cases. Chloroform is used for very young children, and in operations about the mouth and nose which are likely to last some time. Nitrous oxide gas is employed for short operations, such as extraction of a tooth or opening an abscess. Nitrous oxide gas and ether are administered by means of Mr. Clover's apparatus, which is admirably adapted for the purpose. Chloroform is given on lint, from a drop-bottle. Until Junuary, 1875, chloroform was used for nearly all cases ; but I found that, while chloroform frequently depresses the heart's action in long operations, ether stimulates it. It rarely happens that there is any persistent sickness after the inhalation of ether; I have met with but one instance in nearly nine hundred administrations, and in that case it lasted about twelve hours, at the expiration of which time the patient was able to take food well. Ether does not appear to be much less likely to cause vomiting at the time of the operation than chloroform, but it certainly causes much less after-sickness. Chloroform is used in long operations about the mouth and nose, because: 1. The narcosis of chloroform lasts longer than that of ether ; 2. In many operations, as for cleft palate, or removal of the tongue, it is necessary to keep a gag in the mouth, which comes very much in the way of a face-piece, such as is necessary for the administration of ether; while chloroform can very conveniently be given on a piece of lint; 3. In operations for cleft palate, too, there is another objection to the use of ether, in the fact that it excites a flow of viscid saliva, and is apt to induce coughing. In delicate operations about the eye, I prefer chloroform, as it causes less congestion. In cases of fracture which require an anæsthetic whilst the parts are being placed in apposition during the time the muscles are relaxed, chloroform is preferable, because patients recover from its effects quietly, the inhalation of ether being generally followed by a state of noisy delirium and struggling, which would be likely to displace the fractured ends and necessitate their readjustment.

Guy's Hospital.—Mr. Frederic Durham (Surgical Registrar to the Hospital) replies to the queries as follows.—1. In the surgical wards and surgery of Guy's Hospital, chloroform is the anæsthetic which is almost invariably administered; occasionally only, the mixture of alcohol, ether, and chloroform, as recommended by the Committee of the Royal Medical Chirurgical Society, and, more rarely still, ether alone. In some cases anæsthesia first produced by chloroform, is continued by "the mixture" or by ether alone; this especially in cases in which chloroform does not appear to be well borne.—2. Chloroform is given from a piece of lint, fitted into a metal nose-piece for convenience sake alone, a few drops being poured from a stoppered bottle as often as required. The mixture is generally given on flannel, adapted as the loose lining of a cylindrical paste-board or leather inhaler; and ether on a sponge at the bottom of a deep cylindrical leather inhaler fitting closely round the mouth.—3. During the last four or five years, all the anæsthetics, old and new, ether, chloroform, nitrous oxide, bichloride of methylene, the mixture of alcohol, chloroform, and ether, etc., have been used to a very considerable extent at Guy's; but, chloroform being found much the most convenient in administration, and, as a rule, well borne in surgical operations, and much less frequently followed by the disagreeable after-effects—headache, vomiting, etc.,—which were observed to be especially severe and prolonged in the case of ether, it has again become the anæsthetic in common use. In the eye wards, however, "the mixture" is still generally administered.

Mr. Bader (Ophthalmic Surgeon to the Hospital) sends the following replies. 1. For the last two years, I have been using as an anæsthetic a mixture of alcohol one part, chloroform two parts, ether three parts. This mixture I have been using in all cases, both private and in the hospital.—2. It is given in a card-board cylinder, covered with flannel.—3. I have changed the anæsthetic, because I wished to make an experiment.—4. In many cases where the least sign of danger appears, we put from six to ten drops of the nitrite of amyl on a piece of lint, place it on the patient's nose and mouth, with the instantaneous effect of restoring the action of the heart and lungs.

LONDON HOSPITAL.—Mr. Lewis Mackenzie (late resident medical officer) sends the following replies. 1. The anæsthetic chiefly employed at the London Hospital is ether; it was introduced into this hospital in the early part of the year 1872, and has since been used in by very far the majority of cases requiring the production of complete anæsthesia. Some surgeons make exception to ether in certain cases. Mr. Jonathan Hutchinson prefers the use of chloroform in old people with rigid or brittle arteries, as he considers the arterial tension produced by ether, a condition very likely to give rise to cerebral hæmorrhage. It will be remembered that Mr. Hutchinson reported a case of this nature when ether was recently introduced. Chloroform is more used in the maternity department than any other of this hospital; but, in obstetric operations, ether has been chiefly given. Whilst resident accoucheur, I repeatedly used chloroform, and to the full surgical extent in cases of obstructive and operative midwifery. There never seemed any danger; and one has such a powerful reflex stimulus always ready to be acted upon in this class of cases, that it is very rare to hear of any death from chloroform-administration during labour. In a severe case of *post-partum* hæmorrhage, ether seemed to improve circulation, and did not increase the bleeding. In cases requiring the application of the actual cautery about the face, the patient is generally placed under the influence of ether, and anæsthesia is kept up by the use of chloroform. In the eye department, in a few cases, ether has seemed to produce so much venous congestion of the eyeball and orbit that, in some of them (notably cataracts in plethoric old people), chloroform is here preferred. In the dental department of this hospital, nitrous oxide gas has been very freely given, and with great success; but the writer is acquainted with one case in which the inhalation of this gas for tooth extraction was followed by a very painful, intermittent, and irritable action of the heart, extending over a period of one year and more; and again, of another in which auditory vertigo and tinnitus followed the use of this gas, and remained for some months without any cause being discovered by aurists to explain the symptoms. Bichloride of methyline

D

was used for a short time in our ophthalmic department, but was abandoned in consequence of its acquiring a bad reputation in other hospitals.—2. It has been the invariable rule at the London Hospital to have as little apparatus in the administration of anæsthetics as possible : this rule has been maintained because so many gentlemen have to give these agents that it was thought more desirable they should not have each to overcome the difficulties of the various forms of inhalers, etc. ; but should rely on their own judgment and knowledge, and watch for symptoms of danger unfettered. Ether is administered in cones, made with new stiff towels, with sponges in them. Chloroform is always given with a simple Skinner's inhaler ; nitrous oxide gas is given with the usual apparatus ; and when bichloride of methyline was given, it was measured from a drop-bottle on lint. I may mention that it has never been the custom at this hospital to measure the amount of chloroform used. I need hardly add that ether is not measured in quantity before administration ; several leathern inhalers have been tried with the ether, but the towel, in form of a cone, seems still most popular.—3. A great revolution in the administration of anæsthetics took place in this hospital in 1872 ; at that time, ether became almost universally substituted for chloroform. Since the introduction of ether, the hospital has not contained any case in which death has resulted from its use nor during its administration. Cases of death have occurred whilst patients have been under the influence of chloroform ; but none of these could in any sense be classed under the head of " death from chloroform " ; they were such cases as hernia in which the patients were moribund when put on the table, and died during the operation : one was a case of an aneurism bursting whilst the patient was under the influence of chloroform for compression-treatment. It would be quite useless to deny that ether was adopted in this hospital, and readily so, in consequence of several deaths having taken place during the preceding few years from chloroform-administration. I myself, as a student, had the good or bad fortune to see three deaths from chloroform ; two of them, however, were from the introduction of vomited matters into the trachea and bronchi. This raises the important question as to whether it is justifiable to give any anæsthetic when a person is in the last stage of an abdominal obstruction or hernia with fæcal vomiting, etc. Having witnessed several deaths from suffocation by such vomited matters, I should myself be very loath to administer any anæsthetic to patients in this last stage of abdominal collapse. On à priori grounds, ether is a more suitable agent in such cases ; and I personally have never seen death from this cause with ether-administration. It has always appeared to me that, in hospitals, sufficient importance is not paid to the danger of anæsthetics ; fully recognizing and believing that there is never any carelessness in the administration of them, I still hold that often the operation is performed without the precautions which experience should guide us to use. An unseemly haste is a common fault ; everyone seems to want the patient to be under the influence in a minute or two ; the operator looks at the chloroformist as much as to say, " When are you going to get the patient under ? " I think, too often, no examination of viscera by any competent physician is made. Chloroform has been given, and a patient operated on who had acute pericarditis (as was proved post mortem) ; and it teaches its own lesson. Again, how often has chloroform been given without good light in the patient's face, and with an inefficient assistant, when the delay of only a few minutes could have procured good help and sufficient light, etc. Moreover, when possible, galvanic batteries should be at hand, injections of brandy ready, etc. In the case of chloroform, I should attach great importance to the fasting of the patient for four or five hours previously, and the administration of a small quantity of brandy about half an hour before the operation ; for want of the former precaution, several people have suffocated under chloroform. Thus it appears to me that, if we thoroughly examine our patient's " clinical pathology " (as Dr Hughlings Jackson calls it), if we employ the anæsthetic we consider safest (which in ninety-nine times out of a hundred is ether) ; if we commence to place our patient under its influence quietly, cautiously, and surrounded by skilled assistants, and such instruments as galvanism at hand, we shall do all that at present lies in our power; and I very firmly believe that it is only by insisting strongly on the maintenance of these rules, rather than by adding fresh ones to them, that we shall bring the mortality from anæsthetics to its minimum.

CHARING CROSS HOSPITAL.—Mr. Woodhouse Braine (Chloroformist to this hospital and to the Dental Hospital of London) replies as follows. 1. Nitrous oxide is given to render the patient insensible, and the anæsthesia is kept up by means of ether. In some operations about the mouth, chloroform is occasionally given.—2. The ordinary nitrous oxide face-piece is used, and afterwards a felt cone for the ether.—3. I was in the habit of using chloroform. The two anæsthetics now used, when they are fatal, produce death by arresting respiration, and not by paralysing the heart's action; hence artificial respiration in all probability is the best remedy to employ in cases of danger.

At the Dental Hospital, the anæsthetic used is nitrous oxide, followed in long cases by ether.

DENTAL HOSPITAL OF LONDON.—Mr. J. T. Clover (Chloroformist to the Hospital) forwards the following remarks.—1. I am in the habit of using nitrous oxide alone for teeth-extraction and many short operations; and, for longer ones, I sustain the anæsthesia by first giving ether with the gas, and then ether with very little air. This is very satisfactory on the whole, and has the advantage of shortening the period of recovery, as well as that of going to sleep. I think the uncertain amount of chloroform in methylene objectionable. I use chloroform with or without ether in operations on the eye, and those of the face or tongue, or in any operation where it is desirable to diminish the hæmorrhage as much as possible.—2. I give gas and ether by my apparatus for that purpose. The supply of gas is regulated by my foot moving the screw-tap of a condensed gas-bottle. The ether supply is regulated by the hand which holds the face-piece, by turning the stop-cock leading to the ether-vessel. Ether can be given or withdrawn by this means, without removing the inhaler or letting the patient have fresh air. I use ether $P. B.$, of specific gravity .735. For giving chloroform with or without ether, I use a modification of my bellows and bag apparatus.—3. The favourable opinions as to the greater safety of ether, and the increasing alarm as to chloroform, together with improvements in the way of giving ether, have induced me, within the last four or five years, to give ether very much oftener than chloroform.—4. With the view of securing the safety of anæsthetics, I would advise them to be given to the patient fasting for at least four hours. No stimulant should be given by the mouth previously. Chloroform should not be administered to a sitting patient without great care, nor given in proportions strong enough to excite coughing or swallowing. The pulse, as well as respiration, should be watched; and, if there be a pause in the respiration, or if the pulse decidedly lose power, the inhaler should, for at least one inspiration, be taken from the mouth. Raising the chin forcibly away from the sternum should be adopted when, on account of laryngeal obstruction, the inspiratory movement fails to draw in the proper amount of air. A pause in respiration, after the patient has lost consciousness, should be interrupted directly by compressing the chest and abdomen at the same time every two seconds; but if the patient do not respire independantly in less than half a minute, Silvester's method of artificial respiration should be commenced. Ether excites coughing without doing further harm, but should be more diluted after the glottis has become insensible by this or any other anæsthetic.

Mr. G. H. Bailey (Anæsthetist to the Hospital) sends the following replies.—1. The anæsthetics used are: a. nitrous oxide; b. nitrous oxide and ether; c. chloroform; d. chloroform and ether. a. Nitrous oxide is used in all dental operations, and in those of short duration in general surgery.

b. Nitrous oxide and ether are given in all operations, of whatever kind. *c.* Chloroform is used in midwifery and in operations on young children. *d.* Chloroform and ether. Chloroform given in Clover's bag, three and a quarter per cent., with about two per cent. of ether, is best in eye cases, and where the patient requires to be very passive—*e.g.*, in plastic operations, colotomy, hernia, etc.; and, I think, in operations on the throat and mouth, especially as we have usually to keep up anæsthesia with chloroform after administering ether. Further, the great mucous discharge from ether cases is objectionable.—2. Clover's chloroform apparatus, and nitrous oxide and ether-apparatus, are by far the best, and always used by me.—3. Ether, with nitrous oxide, has been used more generally latterly than chloroform, owing, no doubt, to the feeling that ether is far safer than chloroform. The mode of administration of chloroform I have not altered, always having used Clover's bag-apparatus.—4. In chloroform-administration, the patient should always be recumbent, and never more than three and a quarter per cent. of vapour used, with a constant watch, especially on the pulse. Nitrous oxide is better given in the sitting position. For the present, I am inclined generally to give nitrous oxide and ether, as apparently they are safer than chloroform; although, perhaps, we have not yet had full experience on this head.

St. Mary's Hospital.—Mr. H. E. Juler (Chloroformist to the Hospital) sends the following replies.—1. Chloroform is generally used, but in some cases I use the washed methylated sulphuric ether: either the one or the other is employed indiscriminately, as regards the nature of the case. 2. For chloroform, Clover's apparatus is used; for ether, a simple cone of felt, with a sponge placed inside towards the apex. 3. Chloroform has been almost exclusively given here during the last five years, and the ether has only been adopted as an experiment, in deference to its reputation for safety.

Mr. Sydenham J. Knott (late Anæsthetist to the Hospital) forwards the following answers.—1. I am now in the habit of using ether, nitrous oxide, or both combined. Ether I have now used for about two years; and, during the past year, have given hardly anything else for all operations except teeth extraction. And I never give chloroform except when requested to do so by the operator, and that is now very seldom. Since I commenced to give ether, I have become more and more convinced of its good qualities; the chief of these is its markedly lasting tonic and stimulating effect on the heart, as shown by the regularity of the latter, and of the pulse. There is never any faintness, and but seldom vomiting, as compared with chloroform. The way it improved the pulse I particularly noticed when I first commenced to give it; for then I used to give methylene ether, a mixture of methylene, bichloride, and absolute ether. At the specific gravity of 1,000, this fluid has a depressing effect on the pulse; and, as soon as I found the pulse becoming a little weaker, I commenced the ether. Usually the pulse grew stronger than it was before the anæsthetic was employed. Now I generally give the nitrous oxide gas first; this mode is quicker, much pleasanter, and as safe as any known way of keeping a patient perfectly unconscious for any length of time. When I give the methylene first, and then the ether, I adopt a face-piece made of white felt, and covered with black silk, with a small aperture at the small end, and a sponge in the centre of the cone. When I give the gas first, I simply give the gas in the usual way, and then go on with the ether as above stated. The nitrous oxide I give from a hundred-gallon bottle of liquid gas, just outside. I have what I call an exposition jar, the gas passing through which prevents it from rushing along the tube too rapidly to the bag. The chief reasons for my adopting ether in preference to chloroform are these.— 1. It is a much greater stimulant to the heart. 2. It never depresses the heart's action, as chloroform does in some cases. 3. If anything did go wrong with ether, it certainly would not be so sudden as with chloroform. 4. With ether, the cessation of respiration would occur first, and not the sudden stoppage of the heart, as with chloroform. 5. If respiration should cease, there would be a much better chance for the patient than if the heart became paralysed. 6. Ether might be given in almost every case. 7. Fatty degeneration of the heart, aortic regurgitation with hypertrophy, and advanced phthisis with a feeble heart, would make the administration of chloroform a question of grave importance; but in these cases, and especially if the operation were a painful one, and the patient wished for an anæsthetic, I should advise ether. When I give chloroform, I always use Clover's apparatus. The precautions I adopt are, that the stomach should be as nearly empty as possible; that, if false teeth be worn, they should be removed; that no stimulant should be taken just before the administration; that the patient should lie easily and comfortably, and be encouraged to take steady long inspirations.

St. Thomas's Hospital.—Mr. S. Osborn (Surgical Registrar to the Hospital) sends the following replies.—1. The patient is in all cases previously prepared, and no food is administered for four hours previous to the administration of the anæsthetic. Chloroform is the anæsthetic administered to children and old people at the present time; ether is given to adults. The administration of ether to adults sometimes causes attacks of bronchitis; and as it always does so in old people, it is not given to them. Nitrous oxide is also occasionally used, but principally by the dental surgeon. Ether is objected to on account of the longer time required to produce anæsthesia; and chloroform is preferred in all cases except when the patient is very low or subject to chronic disease; and then ether is administered, being considered more safe. When speed is required, chloroform is first administered, and ether is subsequently used.—2. Chloroform is always administered by an instrument known as Millikin's modification of Snow's inhaler. When employed in the wards, it is merely given on a fold of lint. The instrument used for administering ether is Golding Bird's.—3. Millikin's modification of Snow's instrument has been in use since 1856. All changes have been tried within the last three years. For a few times, a mixture of ether, chloroform, and rectified spirits in equal parts was tried, but was discontinued on account of its being very slow in taking effect, whilst the after-effects were unsatisfactory. Clover's bag was tried for a few months, but was discontinued on account of the house-surgeons (there being no special chloroformist) preferring the instrument which had been so long in use, and finding the bag cumbrous. It was found to answer well with women and children, but to be not sufficiently powerful for men, the bag having frequently to be refilled before the operation was complete. Also, on students leaving the hospital, they preferred to be acquainted with an instrument which, when in the country, would be more portable and more easy of application.—4. The subcutaneous injection of one-third of a grain of morphia, one quarter of an hour before the administration of the chloroform, has been found to materially lessen, and in many cases prevent, the occurrence of sickness after administration. Also, the administration of a subcutaneous injection, immediately before commencing chloroform inhalation, has been found of great service where great pain has been expected from the character of the operation. An objection to subcutaneous injections is, that the pupil can no longer be taken as a guide in the administration. When commencing any anæsthetic, I believe the first inhalations should be freely diluted with air, the patient being gradually brought under the influence; and not large quantities suddenly placed over the patient's mouth, sufficient to make him struggle for breath. Chloroform should be always administered, as it is at the hospital, in a small room adjoining the theatre, previously to the patient being brought in for operation, as he does not then become excited, and is more quickly brought under the influence. Great objection should be made to the administration of chloroform in the wards, as frequently I have seen more difficulty in bringing the patients round. Whether this is due solely to their being unprepared, or due to ventilation, I cannot say. It has, besides, a depressing effect on the other patients.